I0489516

Underlying Patterns
By Rick McKeon
Copyright 2015 Rick McKeon

Preface: Join Me On An Adventure!

This is a photographic journey of discovery. We are going to look for patterns in nature and try to discover the underlying forces that produce them. My goal for this project is to keep it simple, thought provoking and FUN! Take a hike in the woods, visit a National Park or the City Park near your home and have fun discovering these patterns!

If you see similar patterns here and there - that's a clue! That's a clue that there are underlying physical principles at work!

I've always been interested in patterns. Early on I took degrees in mathematics and engineering, and throughout my working career my most successful projects depended on spotting patterns and implementing solutions. This book reflects my continued passion for "looking beneath the surface" and understanding unifying forces.

In this book I present a lot more questions than answers. I want to encourage you to discover answers on your own, or at least to propose some theories. That's where the fun lies! It doesn't matter so much if you are right or wrong. What matters is that you are curious and excited to make discoveries. In each chapter I propose a few activities to get you thinking. Each of these activities is highlighted with a little pointing finger ☞ to get your attention. Think of these as your "assignment." They are designed to get you thinking more deeply about the forces that shape our world.

I would like to give a special thanks to my close friend Gerard Coard for his enthusiastic encouragement and support.

Let's get started looking for **Underlying Patterns**!

Table of Contents

1. Very Different but Very Much the Same

1.1 Capture: A Snapshot in Time

Here are a couple of processes captured "in time" by taking a snapshot. Yesterday they were different and tomorrow they will be different, but we have a snapshot of them at this moment in time.

| Icicle | Dried Sap |

These two objects look similar and some of the forces involved are the same. BUT the medium in each case and the processes involved are completely different!

For both examples gravity is pulling downward. In each case they start to assume that classic icicle shape with the top portion wider than the bottom. In one case the medium is water. In the other case it is tree sap. Water is running down the side of that rock face because, at some time in the past, it rained or snowed in the watershed above. Sap is running down the side of the tree because of a wound to the tree. The sap comes from moisture in the ground and up through veins in the tree. The water has solidified into ice because the temperature was below freezing. The sap has solidified because of drying. You can see how different the medium and the process is in each case, but how amazingly similar the results are! They are so different and yet so similar. The underlying forces here seem to transcend source, media and process to produce similar results! This is a hint that there are **Underlying Patterns** at work.

☞ Activity 1.2 Look for Patterns

1. Observe other natural processes that, on the surface, seem to resemble each other. They might be as diverse as the branching of rivers or trees and the blood vessels in your arm. If you see mineral stains on an old wall can they be compared to the spread of a powerful idea throughout a population? Compare the conical shape of an anthill to that of a volcano or a prospector's diggings. How does a complicated freeway system compare to your circulatory system or the ductwork in your attic? Are the same laws of volume and flow involved in all of these diverse phenomena?
2. Are there simple attractors involved? Are there strange attractors involved? What are attractors anyhow?
3. Does it really matter? Does a curiosity about the natural world somehow enhance your life?

2. Conical: An Example of Conservation of Energy?

2.1 Attractors Common to People, Insects and Rocks

Is there a natural law similar to the Conservation of Energy that applies to people, ants, volcanoes and even meteor strikes all in the same way? Is there an attractor toward a conical shape? I have observed ants carrying grains of sand from underground. They carry it right to the rim of the cone and drop it. They don't drop it before they reach the rim because it might roll back, and they don't carry it all the way down the other side – that would be a waste of energy.

For a video showing those ants at work visit my website at rickmckeonNature.com I'm sure it was the same with the old prospectors looking for gold. They needed to remove material from a hole but they weren't going to move it any further than necessary. If you move the material in all directions randomly a cone results.

When I look at the photographs below I have to believe there is an underlying law that "attracts" these diverse behaviors. Animals or men removing material. Pressures exploding hot gas and lava from underground. Meteorites creating an explosion that blows material out from the impact site. Gravity works on all of them requiring more energy to move the material further. Very different phenomena but similar results!

Anthill   Meteor Crater, AZ

Prospector's Test Hole   Tycho crater on the Moon

☞ Activity 2.2 One Law for All Conical Structures?

1. Look for things in nature that are conical and think about their formation.
2. Could the control factors involved be the same?
3. Don't limit yourself to geology. Maybe you see this effect in biological systems or even chemical reactions. Who knows!

3. Waves and Washboards

3.1 Waves Everywhere!

The following wave patterns occur in very different media and are caused by very different forces but they are strikingly similar.

In the first two pictures twigs and pine needles are being washed down an incline. The third is the familiar washboard road that we all love so well. Here the mechanism is not water but car tires pounding the dirt road. The forth picture I call "Waves in the Sky." Here the mechanism is wind acting on water vapor in the atmosphere. The fifth picture shows the pattern that results when moist dirt is frozen, and the sixth one shows waves in a lake. Again the driving force here is wind, but the medium is water instead of pine needles or dirt.

Twigs Washing down a Hillside   Pine Needles on a Trail

| Washboard on a Dirt Road | Waves in the Sky |

| Frozen Dirt | Waves at Lynx Lake |

Each of these patterns seem pretty regular in terms' of amplitude and period. The forces involved are very different: water, wind, physical pounding by car tires and freezing. The materials are very different: loose twigs or pine needles, dirt, water and water vapor. BUT the patterns produced are very similar.

☞ Activity 3.2 Why Do These Patters Emerge?

1. Why are the patterns so regular? Would the washboard road appear different if the dirt were harder, composed of different materials or presented a steeper grade? Would the pine needle pattern appear different if the flow were stronger or the incline steeper? Would the waves different if the wind blew harder over the lake? Does the depth of water make a difference? Would the frozen dirt have a

different pattern if it had more sand and less clay or had a greater moisture content? So, even at a superficial level there are lots of questions to answer!

2. Is it just coincidence that they appear so similar or are they all really controlled by the same underlying force? Could there be one underlying natural law transcending materials and specific forces that produces this particular form?

3. Look for other repetitive patterns – bird song, barking dogs, and group behavior. Are they related in any way to the washboard road?

4. Can you observe these wave patterns in other arenas – psychological, neurological, political, electromagnetic or insect behavior?

4. Fibonacci Numbers in Nature

Some special numbers seem to show up here and there in nature. It makes sense that Fibonacci Numbers would be among them because this set of numbers is based on growth.

It's my belief that if certain numbers show up over and over again, that's a clue. It's a clue that nature has an underlying structure and forces are at work beneath the surface. The theme of this book is to look for superficial similarities and then try and drill down to the fundamentals. In this chapter we are going to look for Fibonacci numbers in the natural world.

4.1. What Are Fibonacci Numbers?

Fibonacci numbers are generated by simply adding the previous two numbers to get the next one. So, starting with 0 and 1, the Fibonacci Sequence is:

0, 1, 1, 2, 3, 5, 8, 13, 21, 34, 55, . . .

See how it works?

4.2. Fibonacci Spirals

Fibonacci numbers appear in many unexpected ways in nature. We don't know exactly why, but because of the way these numbers are generated, they seem to relate to growth. Growth in nature systematically builds on what has come before.

Not all natural things exhibit Fibonacci numbers in their structure, but look at the spiral patterns we find in pinecones, sunflowers and daisies. You will see both clockwise and counterclockwise spirals. In an overwhelming number of cases they seem to be sequential Fibonacci numbers. Here are some beautiful examples to get you thinking.

4.3 White Daisies: CW = 13, CCW = 21

4.4 Pinecones: CW = 8 and CCW = 13

Clockwise Spirals = 8

Counterclockwise Spirals = 13

4.5 Yellow Daisies

On a bright sunny day this last summer I came across some lovely yellow daises. CCW = 8, CW = 13

CCW Spirals = 8

CW Spirals = 13

4.6 Sunflowers

In sunflowers the spirals are especially easy to spot. CCW = 21, CW = 34

4.7 How About Pineapples?

The next time you buy a pineapple count the spirals and see if they are consecutive Fibonacci numbers.

Pineapple CCW   Pineapple CW

Here's a summary of clockwise and counterclockwise spirals I have found.

Object	Clockwise Spirals (CW)	Counterclockwise Spirals (CCW)
White Daisy	13	21
Pinecone	8	13
Yellow Daisy	13	8
Sunflower	34	21

☞ Activity 4.8 Understanding Fibonacci Spirals

1. For the Pinecone and Daisy the number of CCW spirals was greater, but for the Yellow Daisy and Sunflower the number of CW spirals is greater. Will this always be the case?
2. For each species will the numbers be the same or do the numbers get bigger with growth?
3. Can you find other plants or animals with both CW and CCW spirals? How about acorns or cauliflowers?
4. Go out and find some of these things. Look at them closely. Is one set of spirals (CW or CCW) easier to see than the other?
5. In each case we have seen so far, the number of spirals represent two consecutive Fibonacci numbers. Will that always be the case? If so, Why?

4.9 Building a Natural Looking Spiral

Let's "grow" some squares that represent the Fibonacci sequence:

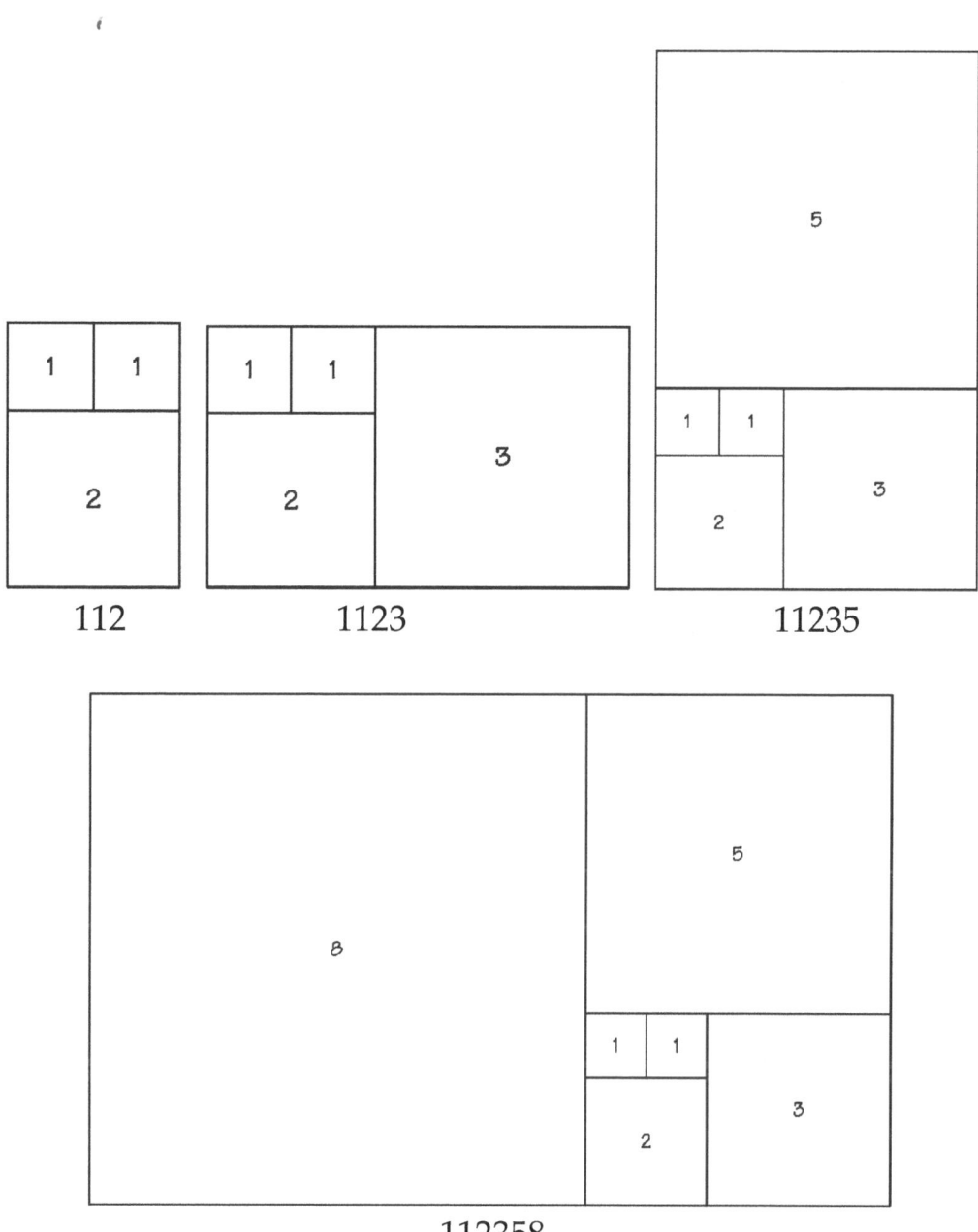

112   1123   11235

112358

Now fit a spiral in this growth pattern.

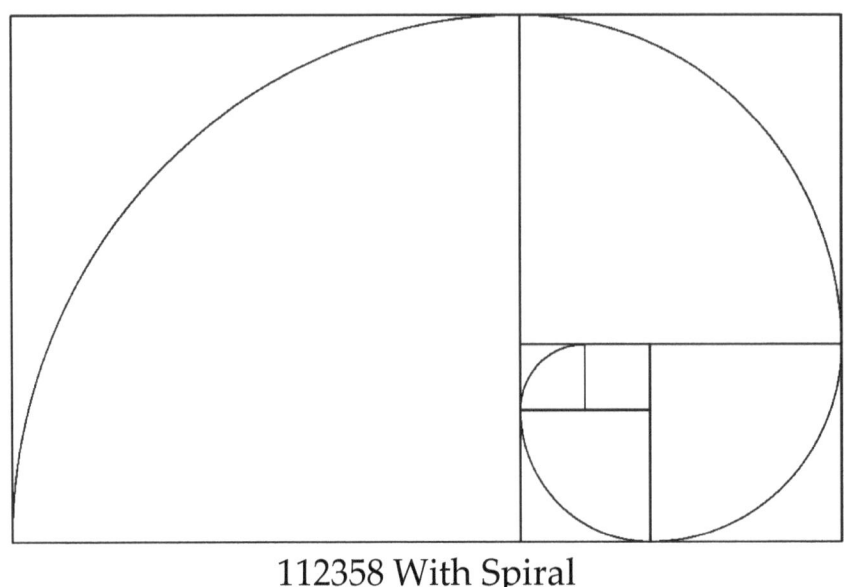

112358 With Spiral

Starting to look familiar?

☞ Activity 4.10 Look for Spirals

1. Think about spirals you have seen in nature. Do any of them have a shape similar to the one we have generated?
2. Think about spiral shells - Snails, Nautilus or Conch.
3. Think about other natural phenomena that take on a spiral form - galaxies, whirlpools in a river, dust devils or even water running down the drain. Even if they don't assume this shape of spiral why do they form spirals at all?

☞ Activity 4.11 Look for Other Examples of Fibonacci Numbers

We have seen a few examples of Fibonacci numbers in nature. Could there be others? Look for things that depend on growth like the branching of rivers or trees, erosion or the growth of animal populations.

5. Fractal Basin Boundaries

5.1 The Watershed Trail

About four miles south of Prescott, AZ on the Senator Highway is a hiking trail called the "Watershed Trail." It gets it's name from the fact that it is on a ridge that divides two major drainages: Bannon Creek to the west which flows into Goldwater Lake, and Sawmill Gulch to the east which eventually feeds Lynx Lake. These two lakes could be called "attractors" because raindrops falling in either basin are attracted to the lake. And the basins surrounding each lake could be called "basins of attraction" for raindrops falling in them. Now, water does flow out of each of these lakes and so those drops are on their way again. For this system we can call each of the lakes a "local minimum." The final destination for those raindrops would be the lowest point they will reach – probably the ocean. That we would call their "global minimum."

Raindrops that fall to the west of the ridge end up in Goldwater Lake, and those that fall to the east end up in Lynx Lake. BUT what about those that fall right on the ridge? Well, that's where it gets a little tricky. Of course the ridge is not a straight line. In fact the closer you look, the more convoluted it gets. This boundary between these basins of attraction is fractal in nature, and so we call it a "fractal basin boundary."

Here's a diagram showing the watershed divide and the basins of attraction toward the lakes.

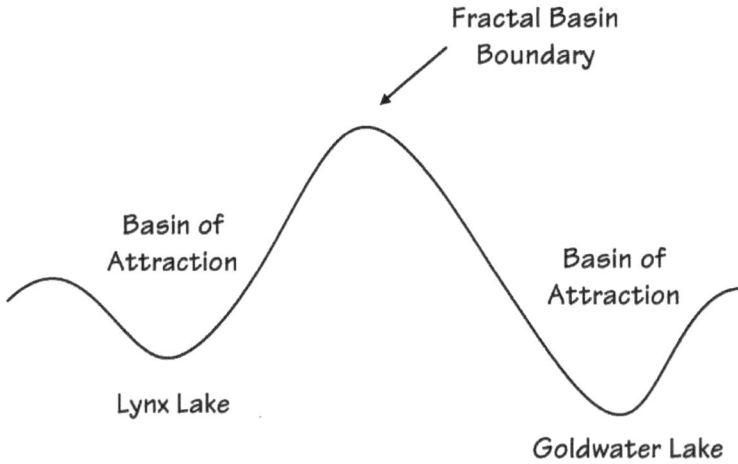

Here's the satellite view showing the watershed divide.

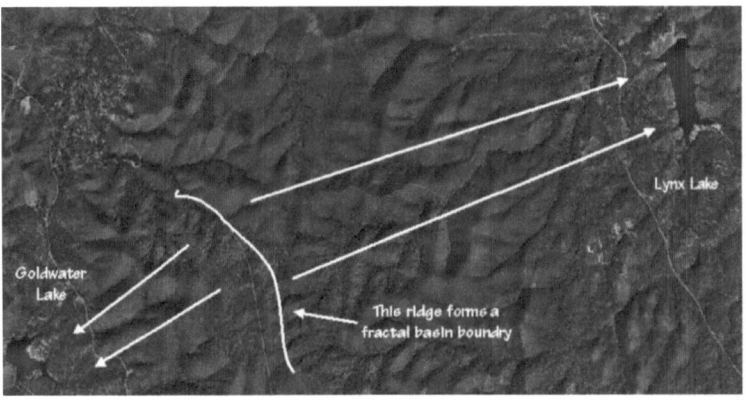

5.2 The Math behind Fractal Basin Boundaries

When equations have more than one solution we can pick a point just anywhere and see if it is attracted to a solution. Some points are pretty close and converge quickly. If we keep track of how many iterations it takes for that point to reach it's destination (like how long will it take a raindrop to reach Lynx Lake) we can assign it a color and produce a beautiful picture that describes the basins of attraction for that equation. Wow! Here's where math and art become one!

To generate the orbit diagrams below I used the program called "Winfract" for an equation with three roots ($x^3 = 1$) in the complex plane. This software uses the Newton method of successive approximations (iteration) to see how long a point takes to get very close to one of the solutions.

For those interested in generating fractals Winfract is a free download. Just do a web search and you will find several download sites. Also, there is an excellent book published by Waite Group Press called *Fractals for Windows: Hands-On Fractal Exploration* ISBN: 1-878739-25-5. This is a very informative tutorial that will get you started quickly.

If we zoom in on the fractal basin boundaries things get really crazy - and interesting! Let's have a look at some color-coded phase space plots and enjoy watching how a point finds its way home.

Here's a look at the basins of attraction and the boundary between them.

Lets zoom in on the basin boundary.

And now zoom in a little closer.

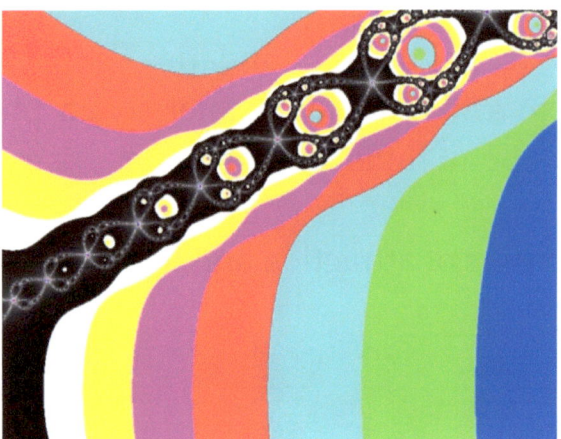

As you can see, it doesn't get simpler as you zoom in – it gets more complex! That's how fractals work!

If you are interested in iteration, have a look at my "Chaos and Fractals" page here: http://rickmckeonScientific.com/chaos.html

☞ Activity 5.3 Look Closely at Ridges and Dividing Lines

If you are out hiking or just thinking about dividing lines, look for fractal basin boundaries. Ask yourself:

1. Does this concept apply right where I'm standing? Can I see specifically where it gets fractal?
2. Can this concept apply to the behavior of animals or even human decision making?

6. Fractal Patterns in Nature

Much of the natural world that we observe all around us appears to be fractal in nature. If you look closely at the pictures below you will notice that these objects look similar at different scales. The overall pattern seems to repeat as we zoom in closer and closer.

Now, there are tons of books devoted to this topic, and I don't intend to simply rehash these ideas. My purpose is to get you excited about looking for fractals in nature and to think about why these patterns appear again and again.

6.1 Fractal Vines

Vines at 10 feet   Vines at 2 feet

Vines at 6 inches   Vines at 2 inches

6.2 Fractal Lichen

Lichen at 5 feet   Lichen at 2 feet

Lichen at 1 foot   Lichen at 4 inches

6.3 Fractal Fern

Iterative processes grew both of these ferns. In one case a mathematical process of iteration was involved. In the other case the instructions contained in the plant's DNA were iterated.

IFS Fern   Natural Fern

0	0	0	.16	0	0	.01
.85	.04	-.04	.85	0	1.6	.85
.2	-.26	.23	.22	0	1.6	.07
-.15	.28	.26	.24	0	.44	.07

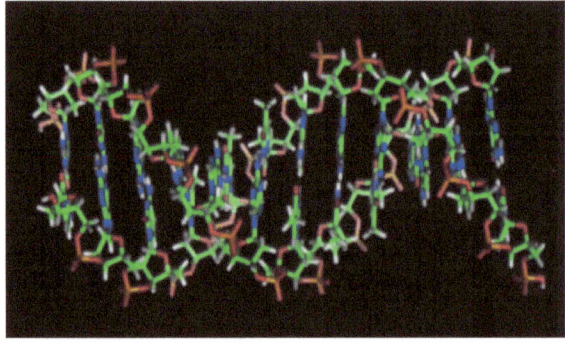

IFS Formula   DNA Formula

Iterated Function Systems (IFS) was developed by Michael Barnsley and is one of the fractal generators in the Winfract program.

☞ Activity 6.4 Look for Fractals

Once you start thinking about fractals you will see them everywhere! Look at various natural objects and see if you can detect self-similarity at different levels. Think about trees or rivers branching, clouds and mountains, patterns of erosion, etc. Look closely and let your imagination run wild!

7. Geometric Forms In Nature

In nature we find many recurring geometric forms. Why do these forms seem to appear over and over again in very different circumstances? Are there underlying forces at work or is it just coincidence?

7.1 Hexagonal Formations

The basalt columns at Devil's Postpile National Monument are very different from a honeycomb. The medium in one case is basalt rock and in the other case it is wax. At Devil's Postpile the hexagonal columns were formed by stress fractures in the cooling basalt. Bees created the hexagonal cells of the honeycomb.

So, the medium and the forces involved in each case are very different, yet the hexagonal pattern is very similar.

Hexagonal Tiling of Basalt   Hexagonal Cells In a Beehive

7.2 Things That Are Round

Things that clump together seem to have a center and statistically only go out to some average radius. What are the limiting factors - heat, nutrition, safety, or is it built into their DNA? These things that clump together may seem very different, but I think there is an underlying principle that unites them. There are many good reasons that humans live in communities. Do some of these same reasons apply to plants and animals?

Lichen   Grass

Cactus   Tree

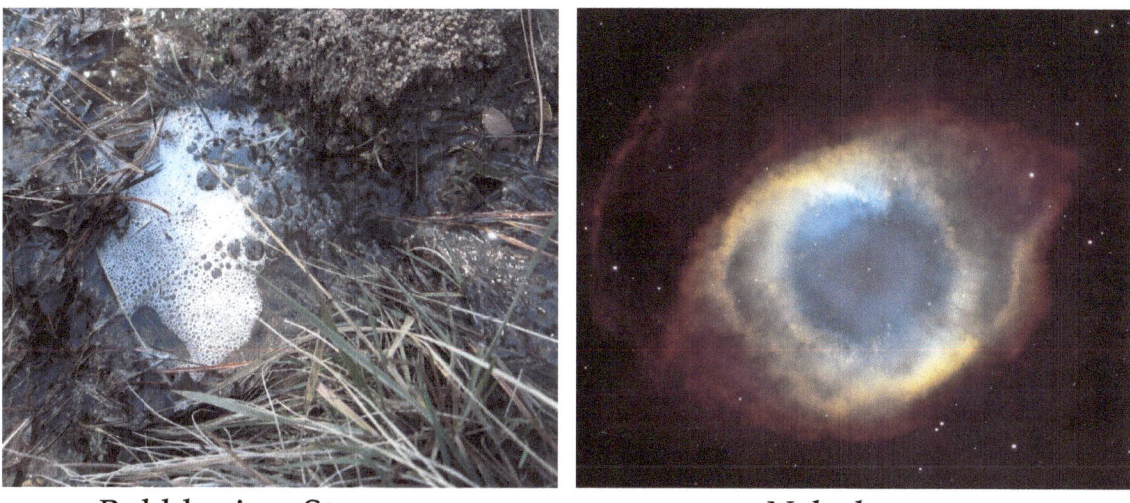

Bubbles in a Stream   Nebula

7.3 Things That Are Spiral

Many natural objects, both living and non-living, have a spiral shape or contain spirals as part of their design. Here are a few examples to get you thinking about spiral objects.

Pinecones   Storm

Spiral Galaxy   Yellow Flower

☞ Activity 7.4 What Creates These Geometric Forms in Nature?

Look for other geometric shapes in nature. Any theories as to why certain geometric forms crop up over and over again? Is there a single underlying force that affects both geologic and living systems?

8. Mounds

Mounds are amazing! People create mounds. Animals create mounds. Geological forces create mounds. Mounds exist because someone or something put them there. It could be a pile of trash or grains of sand brought to the surface by a hot spring. I suspect that they are all somehow related.

The two mounds below look very similar, and they were formed in very similar ways, BUT they were formed by very different mechanisms.

8.1 Anthills

This anthill was formed by ants bringing up grains of sand from under the ground and depositing them in an ever-growing mound.

8.2 Water Anthills

This "water anthill" in the Sierra Nevada Mountains of California near Lake Thomas Edison (a geologically active area) was formed by water bringing up grains of sand from under the ground and depositing them in an ever-growing mound.

☞ Activity 8.3 Look for Mounds

Think about the various mounds you have seen. Maybe you have seen an anthill in the forest, a landfill in your own city or termite hills in Africa. All mounds share certain characteristics, and even if their builders are unaware of it, certain forces are brought to bear in their creation. Look for mounds all around you and try to understand why they are there.

9. Stress Patterns

When people are under stress for a long time they might say, "I fee like I am going to crack." As with people, in nature when stresses build something has to give.

The stress might come in the form of freezing, cooling, drying or mechanical pressure. We are still looking for the underlying forces that unite very disparate phenomena. When mud and rocks display very distinct stress patterns can we look beneath the surface for similarities?

9.1 Various Mechanisms

The mechanisms that produced these stress patterns have gone away but the patterns remain.

Frozen Dirt

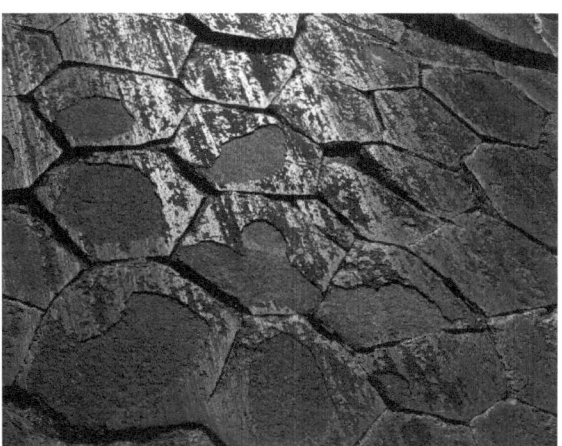
Cooling of Basalt Rock

Pentagonal Stress Pattern   Irregular Tiling of a Mud Flat

At first the tilling shown above might look pretty irregular, but if you count the number of sides for several of them, it appears they usually have just four or five sides. You won't find any with eight or ten sides. Why?

☞ Activity 9.2 What Forms Do Various Stresses Produce?

1. What causes mud to tile this way? Could it be the composition of the dirt, how quickly it dried, how thick it was to start with?
2. Have you seen dirt tile with very large tiles or very small ones? What is the mechanism that constrains their size?
3. Why aren't the tiles found in nature more regular like the tile you would lay down on your kitchen floor or patio?
4. The frozen dirt seems to form a series of straight lines. Why aren't the lines going in a different direction? Why doesn't it form a different pattern?
5. Stress can show up in many different ways. Look for evidence of stress in various materials. How does it change their properties (transmission of light, electrical properties, elasticity)? Are there one or maybe just a few related forces that produce these various phenomena?

10. Turbulence

10.1 Small Scale and Large Scale Turbulence

Turbulence can take place on a very small or a very large scale. What we see on the surface can appear very complex, but there have to be analogous forces at work.

Creek at Thumb Butte Park   Jupiter

Creek Close Up   Jupiter Close Up

Clouds on Earth

10.2 When Does Turbulence Start?

Slow moving fluids can flow smoothly but, as it's speed increases, some critical point is reached where turbulence begins.

☞ Assignment 10.3 Look for Examples of Turbulence in Nature

There have been many models proposed regarding how a smooth flowing stream transitions to a chaotic turbulent flow. There could be several stages of transition from a smooth flow (point attractor) to vortices forming (a limit cycle attractor) to turbulence (Torus attractor or even strange attractor). A deep study of turbulence is beyond the scope of this little book, but if you do some research on turbulence I bet you will find it a fascinating study.

I was sitting on a downed tree that had fallen such that much of it was in a swift flowing river. I could feel the tree responding to the flow and could tell that it was chaotic – not periodic or predictable. Have you had similar experiences with turbulence?

Can you find examples of turbulence on much smaller or much larger scales than the ones shown here? Is there one fundamental law governing all turbulent phenomena?

11. Statistical Distributions in Nature

Have you ever noticed certain patterns in terms of size distribution or spacing of objects in the natural world? This might involve the size of bubbles in a stream, the distribution of pinecones beneath a tree, or even the sound of a waterfall. How about with an approaching rainstorm? You typically hear a few drops, then a few more, etc. before it really starts to pour. It doesn't just go from not any raindrops to pouring instantly. The leading edge of that rainstorm has a statistical distribution of raindrops.

If you can see a pattern, you know these events are not completely random! That means there are some underlying forces at work. Maybe we don't feel or see those forces, but we see the results in the amazing natural world all around us.

We are not going to get involved in an in-depth study of mathematical distributions, standard deviations, bell shaped curves and all that stuff. My goal is simply to get you looking for statistical distributions of things in nature, and to get you wondering about why they happen. Do mathematical laws affect pinecones and waterfalls?

11.1 Here Are a Few Pictures to Get You Thinking

After scrubbing my frying pan for a while I started wondering about the size distribution of soap bubbles. Why just one big one, several medium sized ones and a lot of tiny bubbles? Also, why are there some areas without any bubbles at all? How come not more big ones and fewer tiny ones? The same question could be asked about naturally occurring bubbles in a stream.

A similar question could be asked about the distribution of pinecones around a pine tree. There are very few right up close to the trunk. There are a bunch in a circle around the tree, and then they start to "thin out" at a certain point until there aren't any far away from the tree. How about the distribution in the size of clouds? The picture below shows a beautiful cloud formation. If you study it closely you will notice an area where the size of clouds it pretty consistent, but they start to get smaller and smaller, and eventually merge together. Beautiful and fascinating!

☞ Activity 11.2 Look for Statistical Distributions

1. Listen to a waterfall. If you listen closely you will hear both high pitched and low pitched sounds. Which do you hear more of? What

are the factors that cause this distribution of frequencies? There could be a lot of them. Here are a few to consider: volume of water, height of waterfall, echo or reflected sounds, obstructions in the water, depth of water under the falls, splashing on multiple surfaces or maybe even the wind in the vicinity of the falls.

2. Think about human population densities. There are factors that cause human populations to cluster - resources, convenience, economics or safety. Do animal population densities resemble those of human beings? Are the control factors the same?

3. How strong are the underlying forces or "attractors" here? Are the distributions for this particular phenomenon pretty consistent or do you find wild variations?

4. As usual, I'm asking you to look for common underlying forces. Can a single natural law account for similarities under very different circumstances?

Meet the Author

Hi, I'm Rick McKeon. I am currently living in beautiful Prescott, Arizona. Since retiring I have been spending time pursuing my passion for writing, playing music and teaching. I am currently producing a series of video lessons on playing the banjo and guitar, and am writing books encouraging people to appreciate nature at a deeper level.

Some of my other pursuits include hiking, backpacking, treasure hunting, exploring old ghost towns and mines, recreational mathematics, photography and experimenting with Microcontrollers.

For more about these activities check my other websites at rickmckeonNature.com and rickmckeonScientific.com

Other Books by Rick McKeon

Understanding Nature: Use All of Your Senses to Understand Nature at a Deeper Level!
ISBN: 9781311565129

Sierra Impressions: Images and Inspiration From the Sierras
ISBN: 9781310403699

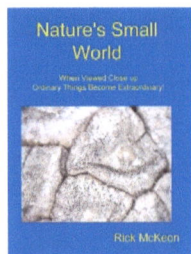

Nature's Small World: When Viewed Close up Ordinary Things Become Extraordinary!
ISBN: 9781507618332

You Make Us Feel Young Again! A Collection of Funny and Inspiring Stories from Rest Home Singalongs
ISBN: 9781310558108

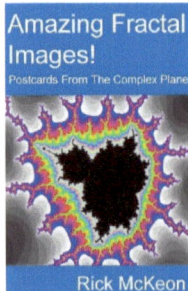

Amazing Fractal Images! Postcards From the Complex Plane
ISBN: 9781311990440

Bibliography

These are some of the wonderful books that have inspired me to look for patterns in nature and the underlying forces at work.

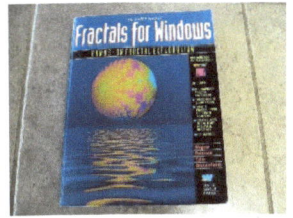 Fractals for Windows, Wegner, Peterson, Tyler and Branderhorst, Waite Group Press, 1992, ISBN: 1-878739-25-5

 Catastrophe Theory, Alexander Woodcock and Monte Davis, Avon Books, 1978, ISBN: 0-380-48397-1

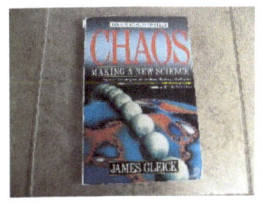 CHAOS: Making a New Science, James Gleick, Penguin Books, 1987, ISBN: 0-14-00.9250-1

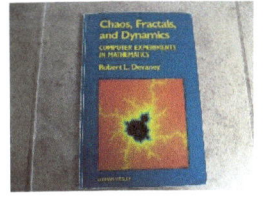 Chaos, Fractals, and Dynamics: Computer Experiments in Mathematics, Robert L. Devaney, Addison-Wesley Publishing Company, 1990, ISBN: 0-201-23288-X

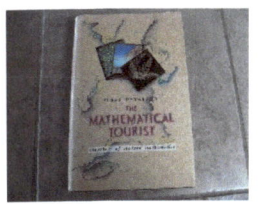 The Mathematical Tourist: Snapshots of Modern Mathematics, Ivars Peterson, W.H. Freeman and Company, 1988, ISBN: 0-7167-1953-3

 Complexity: The Emerging Science at the Edge of Order and Chaos, M. Mitchell Waldrop, Simon & Schuster, 1992, ISBN: 0-671-76789-5

 Simply Complexity: A Clear Guide To Complexity Theory, Neil Johnson, Oneworld Oxford, 2007, ISBN: 978-1-85168-630-8

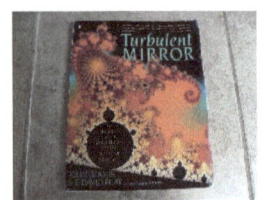 Turbulent Mirror: An Illustrated Guide To Chaos Theory and the Science of Wholeness, John Briggs & F. David Peat, Harper & Row, 1989, ISBN: 0-06-016061-6

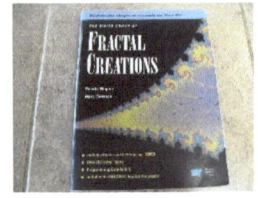 Fractal Creations, Timothy Wegner & Mark Peterson, Waite Group Press, 1991, ISBN: 1-878739-05-0

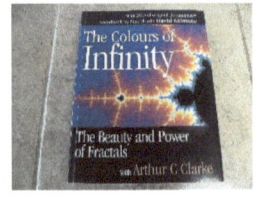 The Colours of Infinity: The Beauty and Power of Fractals, CB Clear Books, Nigel Lesmoir-Gordon, 2004, ISBN: 1-904555-05-5

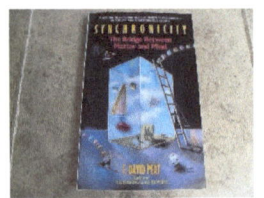 Synchronicity: The Bridge Between Matter and Mind, F. David Peat, Bantam Books, 1987, ISBN: 0-553-34676-8

 SYNC: How Order Emerges From Chaos in the Universe, Nature, and Daily Life, Steven Strogatz, Hyperion, 2003, ISBN: 978-0-7868-8721-7